William W. Ashe

Forest Fires

Their destructive work, causes and prevention

William W. Ashe

Forest Fires
Their destructive work, causes and prevention

ISBN/EAN: 9783337256951

Printed in Europe, USA, Canada, Australia, Japan

Cover: Foto ©berggeist007 / pixelio.de

More available books at **www.hansebooks.com**

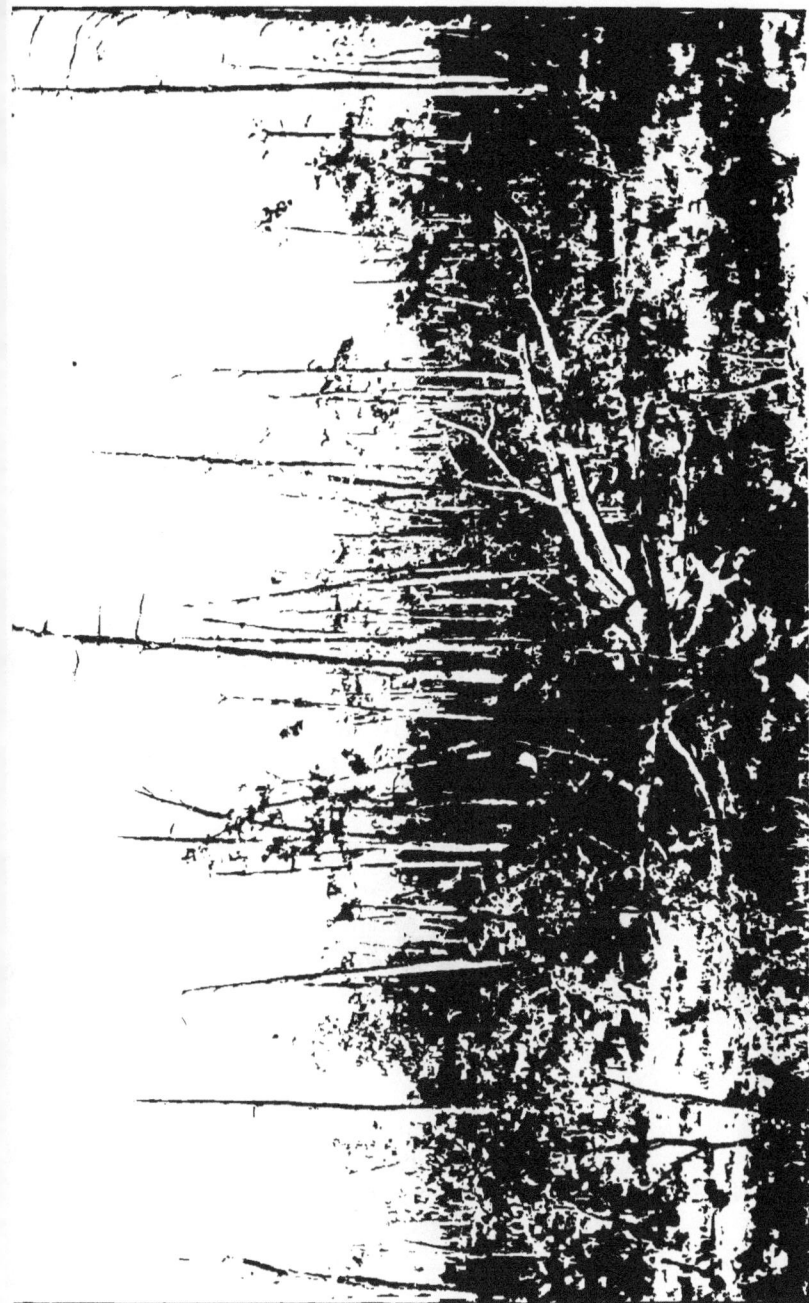

LONG-LEAF PINE FOREST A FEW MONTHS AFTER A FIRE, MOORE COUNTY

Forest fires in North Carolina destroyed more timber than has been harvested

NORTH CAROLINA GEOLOGICAL SURVEY

J. A. HOLMES, STATE GEOLOGIST.

BULLETIN No. 7.

FOREST FIRES: THEIR DESTRUCTIVE WORK, CAUSES AND PREVENTION.

BY

W. W. ASHE,

In Charge of Forest Investigation

RALEIGH:
Josephus Daniels, State Printer and Binder.
PRESSES OF E. M. UZZELL.
1895.

TABLE OF CONTENTS.

CONTENTS.

LETTER OF TRANSMITTAL.

RALEIGH, N. C., January 31, 1895.

To his Excellency, HON. ELIAS CARR,
        Governor of North Carolina.

SIR:—I have the honor to submit herewith for publication as Bulletin No. 7 of the Survey series, a paper by Mr. W. W. Ashe, on the subject of Forest Fires in North Carolina: their destructive work, causes and prevention.

It is earnestly hoped that the people of the State can be induced to give this subject the careful consideration which it deserves, and that such measures can be adopted as will greatly decrease both the number and extent of these fires.

                Yours obediently,
                        J. A. HOLMES,
                            State Geologist.

INTRODUCTION.

The object which it is hoped will be accomplished in the publication of this bulletin is to place in the hands of the more intelligent citizens of the State facts showing the extent to which our forests are being injured by fires, and the importance of adopting such measures as will check or stop this great evil.

Nothing better illustrates the state of public opinion in North Carolina on the subject of forest fires than that the fact that the State law prohibiting such fires, except under certain conditions, has remained on the statute books practically unchanged and unenforced for more than a century. And nothing more forcibly illustrates the destructive work of these fires than the fact that whereas the long-leaf pine, which has for so long a time supplied both lumber and naval store products, a century ago was a common tree throughout the whole of eastern North Carolina, now it is almost unknown north of the Neuse river; and south of this river, at its present rate of destruction, in two decades more it will cease to be a tree of economic importance. Of course the lumbermen have made great inroads on the long-leaf pine forests. But even of the full grown trees the fires have destroyed more than the lumbermen have cut; and so completely have these fires destroyed the young growth of long-leaf pine that in many counties scarcely a specimen can now be found to indicate where once grew valuable forests of this tree.

Among the hardwoods or deciduous trees of the middle and western counties, during the century, the fires have been of incalculable damage to the forests, and in ways that appear to attract but little attention: (1) They have injured the mature trees at their bases, causing them to decay and die; (2) they have destroyed the young growth of the less hardy (but sometimes more valuable) varieties; (3) in case of the more hardy varieties the young growth has been frequently killed to the ground, and the new sprouts from the roots have imperfect bases and make subsequently imperfect trees; (4) these fires destroy the humus on the surface of the soil, and thereby seriously injure its productive capacity.

The injuries resulting from forest fires are thus so manifold and extensive that it is difficult to estimate them accurately on a money basis. But adding to the estimate given on pages 53 and 54 of this report the damage resulting from the destruction of the young growth and the humus, the aggregate damages will doubtless be several million dollars per annum.

Many of the forest fires have accidental origins, such as sparks from locomotives or other engines, hunting parties or camping parties at night, or from burning brush piles during the day. But it is estimated that at least two-thirds of these fires are of intentional origin, and in the majority of such cases the object that is expected to be gained in starting the fires is the improvement of the pasturage. In some cases the result of the burning may be a temporary improvement in that direction, but it must be borne in mind that these fires also destroy much of the grass and annual and perennial herbs and shrubs, by burning both the seeds and the plants, and that in this way, in the long run, even the pasturage in the forest is injured rather than improved by these repeated fires. Unquestionably the wisest policy to adopt, and the policy that has been adopted in all civilized countries, is to provide grass lands for the pasturage of cattle, and protect the forests against fires and all other destructive agencies, because of their value as forests.

The total value of the forest products in North Carolina, as shown on pages 40 and 41, will range from $20,000,000 to $25,-000,000 per annum. The total value of all the cattle in the State is about $5,000,000—a sum which is only between one-fourth and one-fifth of the annual value of the forest products; a sum but little if any greater than the amount of the annual damages to the forests of the State caused by fires.

It must also be borne in mind in this connection that while the total yearly growth of our forests is equivalent to about 9,000,000 cords of wood, the amount of timber now annually cut from these forests is equivalent to 11,000,000 cords. We are thus not only using the interest on this forest investment, but are making rapid inroads on the principal. In the eastern counties we are cutting pine forests to replace which, from the seed, will require from fifty to one hundred years; while in the western counties it will require from one to three centuries to replace the splendid hardwood forests of the rich mountain slopes.

The bare statement of these facts ought to bring us at once to a realization of the truth that the protection and perpetuation of our forest wealth deserve far more consideration than they have received. The right of the present generation to use the mature trees of the existing forests no one will deny; nor will any reasonable man deny that the younger forest growth of to-day should be preserved as the rightful heritage of the next generation. And yet, in the treatment of our timber lands, this latter principle appears to be lost sight of entirely.

It is to be hoped that this matter will be taken up by the press of the State—by the more intelligent citizens of every county and by the lumbermen themselves, who, as much as any other class, are to be benefited by the perpetuation of our forests—and that the agitation may be continued until forest fires are of rare occurrence, and until every sale or lease of timber rights will be conditioned on a rigid protection of the young forest growth against the ax and all other destructive agencies

If such a wise policy is promptly and permanently adopted the forests of the State may continue for centuries to be a source of wealth to her people. If, on the other hand, the present system of waste and destruction continues to prevail, we will not have gone far into the new century before we are brought to a realization of the fact that our splendid forests are a thing of the past.

J. A. HOLMES,
State Geologist.

FOREST FIRES: THEIR DESTRUCTIVE WORK, CAUSES, AND PREVENTION.

By W. W. Ashe.

FOREST FIRES.

The past year (1894), one of considerable drought, witnessed extensive and destructive forest fires in the Northwestern States, involving the loss of over five hundred lives and millions of dollars worth of timber and other property. These fires in the Northwest occurred in a thinly-settled country. But even in eastern Massachusetts, a district dotted with towns, farms and country seats, fires occurred in the latter part of summer, which for several days baffled every attempt toward extinguishing them; and this was not done until a large loss had been incurred. In Massachusetts such fires were unprecedented, for the woodland is carefully protected, and forest fires are only of accidental occurrence. The same cannot be said in respect to other States, least of all for those in which lumbering and the handling of forest products are extensive industries. There is, in most parts of this country, a recklessness amongst those deriving their livelihood from the forest, which it seems, is never satisfied until the source of their maintenance has been destroyed.

The degree of exemption from extensive forest fires which North Carolina has enjoyed during the past year is unusal. In other recent years there have been single fires the losses from which aggregated hundreds of thousands of dollars. Bladen, Cumberland, Harnett and Moore counties have been the seats of such conflagrations since 1890. Although only the large fires attract much attention, the damage and loss occasioned by them does not, on the whole, exceed the destruction caused by the thousands of smaller fires which, each year, pass, almost unnoticed, through the forests. The damage resulting from these smaller fires is inconsiderable, as far as standing timber is concerned and, unless the destruction of other prop-

erty is involved, they are in many localities regarded as beneficial
rather than otherwise. Their damage, however, is great, although
a superficial examination may fail to reveal it. Their continued
repetition means the gradual killing of the forest or the reduction
of it to a few species, which are physically capable of withstanding
scorching heat, which seed or reproduce themselves abundantly and
at an early age, and whose young seedlings are exceedingly hardy.

LOSSES FROM FOREST FIRES IN 1880.

Of all the destructive agencies to which woodland is exposed
none is surer in its effects or more rapid in its execution than fire.
By the census of 1880 the value of the property destroyed by forest
fires in the inhabited parts of the United States, during one year,
was ascertained to be over $25,000,000; and even this immense
amount is supposed to represent only a part of the property
destroyed, since no reports were made of many fires. The same
authority gives the damage from forest fires in North Carolina for
that year as having reached $357,000, and the number of acres
burnt over as amounting to 546,000. The value of the property
destroyed (which was largely standing timber) was equal to that of
all the standing timber which was cut in the State during that year
for saw-logs. And equaling the irreparable damage to standing
timber is the injury to the soil from these frequent burnings. The
harm thus occasioned is rarely even considered.

THE ATTITUDE OF TIMBER OWNERS TOWARDS FOREST UNDERGROWTH.

In circulars sent to well-informed persons in the extreme eastern and western counties a question was inserted concerning the
attitude of timber owners toward the young growth in the forest.
In answering this many correspondents gave prominence to the
fact that it was considered a nuisance to have the young growth in
the forests, and one of the objects of burning, and benefits derived
therefrom, was keeping the thicket of saplings killed down. Where
it is not desired to keep the young growth down in order to secure
a slender pasturage the main object in view appears to be to keep
the woods open for hunting or to facilitate passage through them.

The object chiefly desired, then, in burning is pasturage; the object attained is the destruction of all young growth. While it is regarded as proper and advantageous to secure the maintenance of the mature trees, the utilization of which is possible at any time, and which represent, *prima facie*, a specific market value, no protection is accorded the young growth to enable it to survive. The feeling towards it passes beyond mere disinterest. If the young growth is coniferous it is regarded with disfavor, since it affords no food material for cattle; if it is of broad-leaved trees its legitimate utilization is realized when cattle are browsing upon it. If, in spite of the depredations of live stock, the young growth persists in growing, which is usually the case, it is burned down to effectually check it. The end aimed at in this is to keep the woods open; for in dense woods pasturage grasses do not grow and most broad-leaved seedlings will be shaded out. The pasturage obtained by this means is insignificant, the herbage of many acres being required for the support of a single animal, and it is only secured at the expense of a valuable forest undergrowth.

KINDS OF FOREST FIRES.

The fires in most sections of this State are leaf or grass fires; that is, the dry leaves or dead grass form the fuel which carries the flames. Practically all the fires in the western counties are *leaf fires;* it is seldom that there is sufficient undergrowth present for the fire to be communicated to that, although the undergrowth is usually all killed. In the south-eastern counties, in the long-leaf pine belt, the fires are usually carried by the high and thick grasses which cover the forest soil wherever the woods are at all open. These grasses are hardy perennials like *Aristida stricta. Andropogon tener, A. virginicus, A. elliotti,* and *A. argentius;* the first one known as the wire-grass of the pine barrens, the others as broom grasses or broom-straws or "sedge."

Fire frequently escapes from the grass to the pine trees. In pine woods where the undergrowth is thick, and more rarely in mixed woods, the brush frequently burns and occasions incalculable damage to the standing timber; and in swamps with a heavy undergrowth, particularly peaty ones with a growth of white cedar

(juniper), fires in dry seasons will consume both undergrowth and
brush and mature trees.   These fires of course are the most destructive and are well called *timber fires.*

ULTIMATE RESULTS OF FOREST FIRES.

To ascertain what will be the effects of the continued burning
upon woodland there is no necessity to revert to the condition of
some of the worn-out lands of southern Europe or to cite the
instances from foreign writers of the destitution and barrenness
which have succeeded the continual burning of the forests there.
As concrete evidence of the destructiveness of numerous small
fires as can be adduced is the denuded state of a large part of the
waste sand lands in the south-eastern counties of North Carolina.
The original forests of long-leaf pine in that section have gradually been consumed.   More, it can be safely said, has been burnt
than converted into lumber.   This has seriously affected the supply
of merchantable timber.   Furthermore, these ravages have largely
influenced the supply of the future, in its abundance, quality and
variety, as well as the soil in its ability to sustain a forestal growth.
Already throughout the eastern section of the State profound
changes have been induced in the character of the forest flora of
the uplands and, in places, irreparable damage has been suffered
by the forest lands.   Changes, too, have been produced in the character of the sylva of the hill country and the mountains.   These
changes, however, are not as generally diffused as those which have
taken place in the lower districts, and are less impressive, since
they have influenced the varieties in a mixed hardwood forest,
instead of the struggle for supremacy between two dissimilar pines
or between a pine and an oak growth.   In connection with the
destruction of the long-leaf pine forests it can be said that along
with it is being destroyed one of the most important industries in
the State—that of gathering and manufacturing resinous products.
The aggregate of these products has decreased since 1870 over two-
thirds, with a prospect in the near future of a total failure of the
industry in this State from the inability to secure new and profitable

pineries to tap for turpentine, as the old ones, from long continued boxing, become unprofitable and are abandoned.

THE CHANGES PRODUCED IN THE LONG-LEAF PINE FORESTS BY FIRES.

NORTH OF THE NEUSE RIVER the long-leaf pine formerly extended over the drier sandy land, in an almost unbroken forest, as far north as southern Virginia. This pine has gradually been killed out by fires or removed for commercial purposes, so that but little of the original forest growth now remains. Fortunately there has been a tree ready to take its place, the value of which, for lumber, is but little less than that of the long-leaf pine. This tree is the loblolly pine, which is known throughout the eastern section of this State as *short-leaf, rosemary, old field* or *swamp* pine. Mixed with this second-growth loblolly pine is a vigorous undergrowth of young post, white and red oaks. These, however, do not reach sufficient size to enter into construction, since, after reaching a diameter of fifteen or twenty inches, they generally succumb to internal decay, the soil being unsuited to their growth. This replacement of the long-leaf pine by the loblolly has been so gradual that in places the forest appears, as it stands to-day, to be composed of a mixed growth of these two species. The fact is significant, however, that nowhere is there any young growth of the long-leaf pine. All trees of that species are mature trees, while the loblolly pine is seen in all stages of development from tender seedlings to trees in full maturity.

SOUTH OF THE NEUSE RIVER the effects of the fires have been more marked. The upland soils in this south-east section are of two kinds, the sandy loams which form the level pine lands and the sandy soils which form the pine sand-hills. This sand occurs in a loose layer, from a few inches to many feet in thickness, lying above the more fertile loams. The loams are similar in character to the greater part of the soils found north of the Neuse river, and on them, south as well as north of the Neuse river, the loblolly pine and a heterogeneous growth of small oaks have superseded the long-leaf pine. On the sand-hills the destruction of the long-leaf pine has been accompanied by no aftergrowth of loblolly pine, so that the result of the burnings is large areas of waste land covered only with a scattering growth of the sand black-jack oak, and other scrub oaks, or entirely denuded. The largest tracts of this waste land lie in the eastern counties of New Hanover, Wayne, Duplin, Pender, Onslow, Bladen, Columbus, Harnett, Sampson, Robeson, Brunswick and Moore.

The United States census of 1880 gives the area of the waste lands in these counties to be 105,705 acres. This included only the waste lands attached to farms, as old fields and fallow lands. An examination of these counties undertaken by the North Carolina Geological Survey in 1893 revealed the fact that the area of waste or unproductive forest lands then amounted to over 400,000 acres.

This entire area, amounting to about one-tenth of the acreage of these counties, has been reduced to its present condition almost entirely through the agency of fire. Besides the above, there are extensive areas of thinly timbered pine land far exceeding in extent the area of the waste lands. These are turpentine orchards, either still being worked or abandoned, the trees in which have been continually tapped for turpentine for twenty years or more, and now have their ranks so thinned as to scarcely deserve the name of woodland. The mature trees have one by one been destroyed by fire or prostrated by wind, broken where the collar of the bole was weakened by the deep turpentine boxes. There is nothing remarkable in the fact that the forests have become thinned by the usage of a century, but their openness below and the fact that no young pines are seen taking their place are worthy of the utmost consideration.

The loblolly pine although it quickly, by its winged seed and the frequency of its seeding, gains possession of lands with a moist or loamy soil and attains a medium size, producing a fair grade of timber, is yet incapable of growing on the dry, sandy land which forms a proportionately large part of the area of these south-eastern counties. Indeed if it once succeeds in gaining a foothold on the sand land, where the soil has been cultivated, it is incapable of maintaining there an uninterrupted growth until it becomes a large tree. It is naturally a tree of the lowlands, or rather of the better class of moist or wet loams, and when it does succeed in growing on the sandy lands under such altered conditions it forms only a low growth, with boles short and knotty. Besides rarely ever reaching a height of over fifty feet, it is subject in such situations to "dry-rot" or "red-heart." Associated with the long-leaf pine on this sterile soil are the sand black-jack oak and the highland willow oak, both of them "scrub oaks" incapable of yielding timber, and forming beneath the pines only a low and open growth. The litter of these oaks, with the dry wire-grass and broom-grass of the barrens, forms the fuel for the destructive fires.

CAUSE OF DECADENCE OF THE LONG-LEAF PINE.

As this is a subject of the utmost importance involving the timber growth, or rather the lack of timber growth, on an area of over half a million acres, it may be well to give more fully some of the causes which have, in conjunction with the fires, played important parts in the reduction of these areas to their present state of unproductiveness, and some of the peculiarities of the morphology and life process of the long-leaf pine which prevent it from succeeding in the struggle for existence with its competitors, while they successfully overcome the conditions imposed by man upon forest growth. As a comparison of the differences between the pines of this region and their relations to their environments has been discussed in a previous bulletin of the Survey, only a résumé of the salient points will here be given. These will embody the reason why the long-leaf pine fails while the loblolly pine thrives, even on what was once undebatable ground, entirely within the territory of pure long-leaf pine growth. The fructification, the seed, the seedling, life period and hardiness of the long-leaf pine, considered in their connection with fires, have important bearings on this subject.

THE SEED AND SEEDING OF THE LONG LEAF PINE.

The long-leaf pine, with its fertility impaired by the drain on its vital resources due to the removal of the turpentine by boxing, produces seed abundantly only at irregular and infrequent intervals. A large pine mast occurred in 1845, another in 1872, and in 1892 there was a third, after an interval of twenty years.

In intermediate years the production of mast was small and localized. Since 1892 there has been no long-leaf pine mast at all in eastern North Carolina.

A great proportion of its seed, which are between one-fourth and one-half an inch long, are destroyed by fowls, rodents and hogs, which prefer this to all other kinds of mast. The seed, after falling as the cones open, during October and November and through the winter, are entirely exposed to these varied attacks until they have germinated and securely rooted, which happens in the following spring if the mast is late in falling, or, as is more often the case, if the mast falls in a warm, moist autumn season it germinates at once.

GROWTH OF THE LONG-LEAF PINE.

THE SEEDLING.—The growth of the young seedling of the long-leaf pine is exceedingly slow. The greater part of the vital vigor of the young plant is directed towards the root-system, which consists of an oblong, spindle-shaped tap-root with numerous fine roots diverting from its lower end. The tap-root increases rapidly in length, being fifteen to twenty inches long and one and one-half inches in diameter when the plant is only five years old. The stem, even until the fourth or fifth year, remains nearly stationary in height and only slowly increases in thickness to correspond with the increased diameter of the tap-root. It is during this stage of its development that the greatest number, perhaps as much as nine-tenths, of the young pines are destroyed. Being for so long a time stemless and surrounded by high grass, as is the case in the open woodland or where the merchantable trees have been removed, the young long-leaf pine, on account of its small size and tenderness, is liable to be killed or consumed by fires during a period three or four times as long as the loblolly pine or the scrub oaks in similar situations. A passing fire is apt to destroy nearly every seedling pine in its course. The hogs also destroy a great number of them in this early stage of their growth, by rooting them up to eat their large, succulent roots. Through these two agencies the greater part of the young pine growth is destroyed before it becomes fairly started in life.

LATER STAGES OF GROWTH.—At the termination of the first four or five years of growth, which may be said to form the seedling stage of existence, rapid height-growth begins. During the first few years of this period of height-growth the body axis rapidly develops in length at the expense of thickness, forming long, wand-like stems. During this period of its life wind-storms inflict severe damage upon the young growth, snapping off the slender stems or bending the flexible ones into inextricable masses. Fires, too, are still capable of doing great damage, since the loose, thin bark affords but slight protection to the living wood against heat. The resulting damage, however, will be most severe from spring fires, which occur when the new wood is forming and the leaves shooting. After a height of twenty to twenty-five feet is reached the height-growth is accompanied by rapid increase in diameter, and throughout the rest of the life period neither height nor diameter-growth alone takes place at the expense of proportional development elsewhere.

On the better class of soils at the expiration of twenty-five or thirty years the trees have attained a height of thirty to thirty-five feet and a proportionate diameter of ten to twelve inches. The rate of height-growth probably reaches its maximum before the twentieth year and gradually decreases after that time: being rapid from about the twentieth to the fiftieth year, averaging from eight to ten inches a year; being moderate from the fiftieth to the eightieth year, averaging six to eight inches a year in height; after that period being slow, not averaging over two to four inches in height-growth during each growing season.

The diameter-growth is slow until about the fifteenth to twentieth year and extremely rapid after that until the fortieth to fiftieth year, when it becomes considerably slower. These variations in the manner and rate of growth apply only to forest trees; that is, trees grown in large bodies with a thick stand, so as to fully shade the ground and crowd and overtop each other as soon as height-growth is fairly under way. While these laws of the rate of accretion, or increase in volume of wood, are deduced from the rate of growth of what is called old-field growth, where growth began simultaneously with all trees, they are probably applicable to isolated young pines appearing in a hardwood growth or to young pines springing up in a pine forest where the mature trees are thick. Entirely different laws, however, govern the rate of growth of isolated specimens, or where the young trees are far apart: in which cases from the start there is rapid diameter-growth and crown development at the expense of height-growth and clear, straight stems.

In the boxed trees which now constitute so large a part of the long-leaf pine forests, practically all of the merchantable timber having been boxed, the rate of growth is exceedingly slow, ten or twelve years being required for one inch in diameter to be gained, and the annual increase in height not being over one to two inches.

The long-leaf pine in the forest forms no appreciable quantity of wood between the first and sixth years, and very little between the sixth and fifteenth years. During the next thirty years it adds rapidly to its bulk, and then begins a gradual decrease in the volume of wood added, which decrease continues normally at a uniform rate until old age destroys the specimen.

VALUE OF DETERMINING THE RATE OF GROWTH.

While a small tree, ten to twelve inches in diameter and all sapwood, serves admirably for the manufacture of resinous products, it is only after a very much longer period of growth that a tree reaches a size suitable for saw-logs. This period is dependent, of course, on the rapidity of growth, which varies some with difference of soil. If heart timber is desired, and this is the only kind which is in demand so far as the long-leaf pine is concerned, the time required for a forest crop to reach maturity, when the trees have attained a size large enough for applying them to all uses, is certainly not less than one hundred years. This pine is about twenty-five years old before the heart-wood begins to form, and after that time, speaking in a general way, a ring of heart-wood is formed internally for every ring of sap-wood added to the periphery. So that to estimate the time required for the growth of trees with the heart-wood of a specified diameter the twenty-five peripheral rings of sap-wood must be taken into account, as well as the decrease in the rate of diameter-growth after the fortieth or fiftieth year.

The accumulated wood in the long-leaf pine trees, which form the forests as they stand to-day, represents the results of not less than one hundred and fifty years of uninterrupted growth and probably more than one hundred and seventy-five years; that is, the trees range in age from one hundred to two hundred and fifty years, while all of the trees less than one hundred years old, or about two-thirds of the forest body, have been killed out by fires or prevented by them from developing.

SUMMARY.

The infrequency of the seeding of the long-leaf pine enables trees which seed more frequently to prepossess the ground. Its slow growth renders it more liable to be killed in the tender, seedling stage. Its seed as well as the roots are eaten by hogs. Its slender stem makes it more easily broken off during the sapling stage. It does not bear seed until five or six years later than the loblolly pine of the same age. In all stages of growth, on account of the thin bark, it is exceedingly sensitive to fire and heat, which, without burning them at all, frequently kill large trees. When a tree is but partially scorched the borers attack it and finally kill it. Vigorous trees between eighteen and thirty years old, which have a thick, firm bark, appear to be least injured by spring fires.

DESTRUCTION OF PINE TIMBER BY FOREST FIRES.

In spite of the hogs, if fires had been kept off of these long-leaf pine lands during the past sixty years, or since the extensive tapping for resin began (about 1835), the forest density would now probably be no less than it was before that period. The trees which are now being removed for lumber could still have been taken and with good results; for those left would have grown with renewed vigor as the thinning was carried on. It appears that at least one-half of the standing timber has been destroyed by fire and the loss to the people from this item alone has been immense. The low price of long-leaf pine lands is a source of constant dissatisfaction to its owners, while the truth of the matter is that at least one-half of the merchantable timber has been burnt off, and

smaller trees which would have become merchantable timber have
been prevented from doing so by the forest fires.

FOREST FIRES A CAUSE OF THE DECREASE IN NAVAL STORE PRODUCTS.

For the past eighteen years there has been a constant decrease
in the output of naval store products in this State. This is directly
due to the fact that the existing orchards are becoming exhausted.
Since 1830 the fluctuations of the exports of resinous products
from Wilmington have been indices of the prosperity of the industry in the entire State; so that to say that the exports of rosin from
Wilmington fell from 700,000 barrels in 1870 to 450,000 barrels in
1880, to 385,000 in 1890, and to 200,000 in 1894, merely expresses
the rate of the decline in production for the State. An examination of the lands of the pine belt completed in the spring of 1894
disclosed the fact that there were, at that time, only 55,000 acres
of original growth long-leaf pine timber which had not been
boxed; and only 33,000 acres of young growth long-leaf pine
which had become trees, or which had reached a sufficient size to
insure their becoming of commercial value; and some of this had
already been tapped for turpentine. As has been stated before,
this young growth lies almost entirely in what were at one time
old fields, or enclosures by which it was protected from fires and
hogs. Only in New Hanover county has any considerable amount
of young growth been observed on forest lands; and there a law
prohibiting stock from roaming at large through the forest lands
has been in force for a number of years.

When the producer of turpentine burns his woods to keep them
open and facilitate the collection of turpentine he fails to consider
that all the young growth, from the youngest seedling to the mature
stock, represents so much invested capital; and that the fully mature
trees alone represent the accrued dividends, which ought to be so
utilized as in no way to interfere with the great productive portion
of the capital. In neglecting this the productive body has been
destroyed as worthless, while the mature trees, which represented
only one crop evolved from it, were held as the only prize.

However deleterious to the mature trees may be the system of
collecting resin which is in use throughout the United States, its

destructive results are insignificant as compared with those produced by using fire to keep down the "troublesome growth" of young pines. The entire forest reserve has been made subservient to the means used to lighten the labors of those engaged in the collection of these forests by-products.

THE EXHAUSTION OF THE LONG LEAF PINE FORESTS.

The exhaustion of the long-leaf pine forests of this State may be looked for now at no distant date. The waning turpentine industry will hold its own, as an industry of some importance, possibly for ten years longer, though there will certainly be a steady decrease in its yield each year. The production has possibly been stimulated during the past two years by the fact that while the agricultural products of this section were depressed in price the prices for the better grades of resinous products have, on the whole, been more favorable and the demand firm. With a general rise in the prices of farm products the working of many turpentine orchards would probably be relinquished.

The long-leaf pine lumber industry has probably reached its maximum expansion in this State, and its decadence when once set in will be rapid. The largest bodies of compact forest are in the counties of Moore, northern Richmond and eastern Montgomery. These will last for ten years if the drain upon them continues at the present rate and there are no large fires; but there probably will be many fires. Two of the largest fires since 1890 have occurred in this district, consuming or damaging timber to the value of over $200,000.

INCREASE IN THE AREA OF WASTE LONG—LEAF PINE LANDS.

Including both the forest lands and the old fields and fallows, there are now at least 600,000 acres of waste or unproductive land. This area, as large as it is, may be expected to increase during the next decade to about 1,000,000 acres, as lumbering proceeds on the sand-hills, where no loblolly pine succeeds the long-leaf pine. About 75,000 acres of this sandy land are yearly denuded to furnish the output of 150,000,000 feet, board measure, of yellow pine lum-

ber; and the greater part of the land, having served what is regarded
as its final use, is left uncared for and kept, by the continual burnings, idle and unproductive, covered only with scrub oaks which
withstand the fires.

THE NEED TO PROMPTLY STOP FIRING THE PINE BARRENS.

In 1892 there was an abundant and generally distributed long-
leaf pine mast. The young seedlings from this were to be seen
everywhere the next spring, but a large part of them have already
been destroyed. Even on lands which have been lumbered, severely
burnt or worked for turpentine for a long period and then abandoned
a good many years ago, and which had only one or two mature old
seed-bearing trees of long-leaf pine to the acre, there were in the
spring of 1893, following this heavy long-leaf pine mast, hundreds of young seedlings to be seen scattered among the young
scrub oaks or beneath the old ones. Given half a chance and they
would all have grown. Many of them are still left; and no greater
benefit could be conferred upon the south-eastern section of this
State than to have these pines protected, and allow them to grow.

Although these south-eastern counties have for so long supplied
such a large amount of high-grade timber, it is nevertheless a fact
that in many places good timber, for domestic uses, is becoming
scarce. With all the timber being cut for shipment and none
reserved for home use in the future, and all the young growth
which would in time make timber being killed by the fires, there
is no reserve timber which can be relied upon to meet the demands
even of the next few decades. If these young pines are all killed
out, it means just so much loss to the people; and it is the people,
individually and as a body, who should take steps to prevent this
result. No effort should be spared by them to protect the young
seedlings where any are yet to be found and to see that they are
enabled to grow into mature trees.

It is only by so doing that the timber supply of this region, even
in the near future, can be made a certainty. It will probably be
many years before there will again be a mast of the long-leaf pine
equaling the recent one, either in distribution or in abundance, and

in the meanwhile a large number of the seed-bearing trees in large bodies, or the stunted trees left in lumbering, or the scattered trees in the turpentine orchard, will have been destroyed, so that neither the aggregate number of the young seedlings nor the uniform thickness of the stand can possibly be so great again in the near future as at the present time. It is no more difficult to give the necessary protection now than it would be at a future date, while the benefits derived from prompt action and careful watching will now be manifold. No one, who has at all investigated the situation, will deny that the young long-leaf pine growth must at some time be afforded protection, if there is ever to be a regrowth of lumber trees in this section on the sandy soils.

EFFECTS OF BURNING ON PINES, CYPRESS, CEDARS AND BALSAMS.

The effects of fires on the long-leaf pine seedling have already been stated. One and two-year-old loblolly and short-leaf pines will sometimes shoot from the roots if burnt down; and these trees by reason of their rapid growth and thick bark are little injured by ordinary fires after they have reached a height of fifteen feet and a diameter of four inches; the large trees of both of these species have thick and close barks. The pocosin pine, like the pitch pine (the black pine of the piedmont counties), has remarkable recupera-tive powers, young seedlings when burnt down for several consecutive years still continuing to sprout from the stump or roots, and fair-sized saplings when killed above put up root sprouts. These root sprouts, however, do not make large or vigorous trees. White pine seedlings growing on dry ridges with oaks are very tender and do not put up again after being killed down.

Cypress swamps, even in very dry seasons, are rarely burnt over, and although when this happens the damage to large trees is usually small that to the younger trees and shrubs is considerable. Small cypress, however, if under three inches in diameter, will frequently sprout from the stump. Neither red cedar nor white cedar (juniper) probably sprout again after having been burnt down. Oaks or pines often take the place of a growth of red cedar; and the white and sweet bays and gums will take the place of a growth of white cedar if the soil is not too peaty for their growth. The forests of black spruce and Carolina balsam along the summits of the highest mountains are sometimes burned. The seedlings of neither species sprout again. A growth of wild red cherry (peruvian) mixed with some black cherry and yellow birch forms the aftergrowth, which is in time supplanted by the spruce and firs again. The hemlock-spruce (spruce-pine), owing to the damp places of its growth, is rarely injured by fire. Neither it, however, nor the Carolina hemlock sprout from the stump or roots after the stems have been cut or burnt.

BROAD—LEAVED OR DECIDUOUS—LEAVED TREES.

The oaks vary a great deal in their behavior after a fire. Nearly all of them, if young and vigorous, will sprout from the roots after the top has been injured. While this is true, their sprouts, like those of most other trees under similar conditions, are never certain to make large or sound trees. As the dead stock decays an opening or scar is apt to be left at the foot of the tree, in which rot is almost certain to begin sooner or later. This seems to be more certain in the case of black and red oaks than in the case of white oaks, which latter, it seems, are attacked by fewer rot-causing diseases than the red oaks.

But few of the oaks sucker if the burning is repeated many times in succession. While the trees from shoots are usually unsound, those from suckers are decidedly short-lived and are much affected by drought, owing to the fact that their root systems are superficial and they have no tap-roots or deep-seated lateral roots. This is well known to be the case with apple stock in which the grafts have been made on suckers instead of upon seedlings. The oaks which sucker most seem to grow on sandy or light soils; the live oak, the laurel-leaved oak, sometimes the post oak and sometimes the high-ground willow oak and the water oak. Young trees of the Spanish, black and scrub oaks are capable of withstanding a fire which would kill outright equal-sized trees of the white and, probably, of the post oaks. Some of these trees, as the black-jack, seem to be, as far as an ordinary leaf or brush fire is concerned, nearly fire-proof. Sapling red oaks, too, in the higher mountains are very hardy.

While all of the hickories and the white and black walnuts are extremely sensitive towards fire, they all sprout more or less from the root or collar when young. Some of them, as the white hickory, will even sprout, when six to eight inches in diameter, if killed down in the early spring when the sap is rising, though if killed later in summer they will not sprout at all. The birches all sprout, some from the roots, while sapling beeches if top-killed will frequently sucker besides. Young sour-wood trees sprout freely if killed down. Black gums are hardy and even when young are not much injured; but while sweet gums are more injured they put up vigorous root shoots. The Ailanthus ("tree of heaven"), which has in many places made itself at home in the woods, and which affords a valuable timber, suckers so as to form dense thickets of young shoots, even when trees twelve to eighteen inches in diameter are top-killed by fire; and especially is this so if the trees are in the open or under a light cover. The white bay in the lower district suckers from the constant burning, so as to form dense clumps five to ten yards in diameter of low, scrubby trees; and in places, but more rarely, the same is true of the yellow locust in the mountains. The chestnut is one of the few useful trees, which is not only fairly hardy, but also when killed down sprouts vigorously and abundantly, and its sprouts reach a large size. But even it cannot continually propagate itself in this manner, and unless the woods are renewed by seedlings with fresh life and vigorous roots the chestnut, too, will begin to become scarce.

All of the trees which are especially worthy of protection, on account of their economic value, are greatly injured by fires. But some of the most valuable hardwoods, as the white oak, hickories, chestnut and locust, succeed in securing a stand after light fires.

Leaf fires in the late spring burn up the seed of the elms, as they ripen and drop their seed in March and April. Fires during the winter and early spring are apt to destroy the germinating powers of most seed then on the ground. The seed of several of the hickories, the more valuable ones, the walnuts, the buckeye and some of the oaks can resist considerable heat. A few trees, such as the soft maples and the box elder, ripen their seed in summer and as they sprout at once there is no chance for them to be

2

destroyed before at least making a start. The scorching leaf fire, though, destroys thousands of seed in every acre of woodland through which it passes.

EFFECT OF BURNING AS SEEN IN THE COMPOSITION OF THE FOREST.

If the fires are merely leaf-fires, as are most of the fires in the hill country of this State, but are regularly repeated, the seedlings of tender kinds are kept killed down, while those of kinds which do not sprout from the root are killed outright. This in the end, after repeated burnings, will make the forest composed almost entirely of trees which sprout from the stump or roots, and only the hardiest of these will survive, no matter whether they may be valuable or worthless. This is seen in the way the hardier oak growth has taken the place of the pines in many places in the eastern section of the State; for when protected from fires, as in old fields and in enclosures, the aftergrowth is almost of unmixed pines.

In the middle section the hardier black and black-jack oaks are taking the place of the hickories and, in places, of the white and post oaks, while the young growth of the tulip poplar has entirely disappeared from over certain areas, or is confined to the moist lands along streams. In the upper districts the sourwood, from its rapid growth and readiness of sprouting, has become exceedingly abundant, to the exclusion of more valuable trees. In the lower mountains the scarlet and black oaks have been enabled to greatly increase their numbers from their hardiness, while at higher elevations, where the chestnut is common, there are found groves of open chestnut woods, the trees short-bodied and broad-crowned, where there was at one time a mixed forest of red and chestnut oaks, and other trees. Here the chestnut, from its power of producing vigorous shoots, has alone been able to maintain itself, while other trees, which are accounted unusually hardy, have largely succumbed. Such groves are not infrequent, on southern slopes, through the forests of the Appalachian plateau, between 2,500 and 4,500 feet elevation. Few young trees are to be seen anywhere among the chestnuts. The red and chestnut oaks and the chestnut are the most common trees on southern hill-sides and slopes in the higher mountains. Everywhere they are seen in varying proportions, according as they have been injured by fire, in which case the chestnuts outnumber the other species; or the nuts eaten by hogs, when the young red oaks become the more numerous, since the fastidious hog prefers the sweet nuts of the chestnut and the chestnut oak to the bitter acorn of the red oak. But slight damage is wrought by fire to the forests of the higher mountains on northern slopes, owing to their openness, the dampness of their humus, and their never having been lumbered or extensively culled, and hence not being encumbered with lops or thickets which would furnish fuel for fires.

FIRES CAUSE DECAY AND HOLLOWS AT BASES OF FOREST TREES.

The greater number of the hollow, immature and undersized chestnuts, yellow poplars and oaks in the middle and western counties were formed either directly or indirectly through fires. Fires scorch small spots near the bases of trees, which cause the death of the liber or generative tissue from which the tree increases in diameter. The bark then becomes loosened, or falls off, over the spot, and the tree presents on its surface a layer of dead sap-wood. The sap-wood of most trees is, as is well known, more subject to the attacks of insects, borers, beetles, etc., and to rot than is their heart-wood. If the bark is merely loosened insects will quickly enter the spot, and will channel sometimes into the heart-wood. Through their openings air and water afterwards enter, leaving the germs of decay. In this way the entire interior of the base of a tree may be made hollow and, the bark remaining intact, exhibit no signs of the rottenness which is within.

If the bark falls off immediately from the burning the dead sap-wood is then directly exposed to rot, and fungi in the shape of punk, or mushroom-like excrescences, grow upon it. These and their representatives in lower and more obscure forms are the causes of rot. Their roots, as their vegetative parts may here be called, in the shape of minute white threads, penetrate the dead wood, at first the dead sap-wood and then the heart, and break down the structure of the wood, so that finally great hollows are left. Successive fires, too, will gradually eat into the dry and deadened wood, and not only burn out the heart-wood but consume

the walls of sap-wood.   The base of the stock is finally greatly
weakened, and storms prostrate thousands of such specimens whose
stems above the hollow buts are perfectly sound.   Or if they stand,
the rot and the fire will gradually destroy so large a portion of the
cylinder of heart-wood that the value of the tree for lumber will be
greatly lessened.   All such, however, will finally windfall.

EFFECTS OF BURNING ON THE SOILS OF FORESTS.

Burnings probably produce few changes in the inorganic parts
of soil; at least, such as may be produced, are unimportant.   But
all the organic ingredients which lay on the surface are destroyed;
and peaty soils, which have a large proportion of inorganic material intermixed throughout their whole structure, are often seriously damaged by having this organic matter consumed so as to
have their industrial value lessened or destroyed.

HUMUS AND ITS FUNCTIONS

One of the most important requirements of forest growth is the
humus, which is the accumulation beneath the trees of fresh and
decaying organic matter, mostly shed by the trees themselves in
the shape of leaves, or by the undergrowth beneath the trees.

While humus, of course, cannot take the place of inorganic soil,
it largely supplements it in its chemical relations, and besides per-
forms important physical functions in connection with tree life.
Its depth and richness depend to some extent on the character of
the soil and the kind of growth which it supports.   It is deeper
usually under broad-leaved trees than under conifers, and under
trees like beach and maple than under oaks.

For the trees of North Carolina its depth and the rapidity of its
accumulation are almost directly dependent upon the shade-bear-
ing power of the trees producing it.   There are a few exceptions to
this, as the black spruce, which, though shade-enduring, forms only
a poor humus, and the tulip poplar, which, though light-enduring,
forms a humus of good quality.

HUMUS A RESERVOIR FOR WATER.—Humus is exceedingly hydro-
scopic.   The interfibrous capillary meshes and net-work absorb
water, or even moisture, with the readiness of a sponge.   During a

rain it imbibes water until saturated, which adds greatly to its value on hill-sides, where, if the ground were naked, all the water which fell, and which the soil did not at once absorb, would drain off superficially. In the humus, or mold, it is stored away and gradually given up to the soil beneath, as that becomes drier through evaporation from the trees or through subsoil drainage. The importance of preserving this humus is seen, too, in the fact that it equalizes the flow of water in our rivers and streams, and, besides holding in check the great freshets, enables more water to be utilized in dry seasons by the hundreds of cotton, woolen and grist mills in the State which depend upon water for their motive power.

The evaporation from compact humus is less than from soil in similar situations, since the heat-conducting power of humus is less. Every agriculturist realizes the value of this fact, and utilizes forest litter as a mulch for retaining moisture in the soil. To the sylviculturist, or grower of trees, this property of humus is of especial importance. One of the paramount requisites for tree growth is moisture. Indeed, next to the physical properties of soil none other of its qualities seems to be more influential in determining the character of the forest growth than the proportion of moisture present.

Another property of humus mentioned above is that it is a poor conductor of heat. For this reason it is affected by sudden thermal changes less readily than drier soil is. The value of humus to woodland in this relation was fully exemplified during the spring of 1894, at the time of the sudden and late frost. The severe frost of April, besides destroying the foliage and nuts of all white hickories, which are the most abundant and valuable hickories in the middle district of North Carolina, killed hundreds of these trees. The trees killed stood on southern hill-sides or crests, where the ground was open and there was no humus, and these trees had partially leaved out; while no or only slight injury was done to other hickories of the same kind which were situated with like aspect but in thick growth, and with a good humus around them, and whose buds consequently, being less effected by the spring warmth, had not yet exfoliated. This method of mulching with humus to delay foliation or florescence, as a safeguard against late frosts, is artificially practiced by orchardists, especially peach growers. The hickories killed by this freeze were almost entirely mature trees. the young ones being more usually found where there is considerable undergrowth, and consequently some humus.

THE BENEFICIAL CHANGES HUMUS UNDERGOES.—The chemical properties of humus are regarded by agricultural chemists as being of the first importance in farming, and they lose none of their value by transference to the forest. The chemical elements found in forest humus are, of course, those necessary for the building of trees, since the humus is directly derived from this source. The natural changes which this humus undergoes are such as are most beneficial for the growth of the surrounding forest, and show the power of organic life to husband its resources and render available for its own nutrition the products derived from the disorganization of its own wastes.

Humus, by its decay, forms organic acids, and these in turn form salts by combining with the mineral bases, potash, lime, soda and magnesia, which exist as such in the leaf tissues. These salts, like the humus itself, are only slightly soluble in water, and this prevents them from being entirely washed out by drainage, and their usefulness, as far as the soil of that locality is concerned, being forever lost.

Sooner or later, however, these compounds undergo another change, combining with stronger acids of the soil, when their fixation is complete, and they return to the original or similar compounds from which they were primarily derived by plant life.

Ammonia, or nitrogen, in the form of compounds available to the living plant, is one of the expensive constituents of many commercial fertilizers, and is one of the bases of their value. The nitrogenous waste from humus is in the form of ammonia combined with some of the plant acids, or other organic compounds, or with part or all of these compounds replaced by a mineral acid. Finally most of it becomes fixed in the soil as nitrates of alkalies, lime, magnesia, potash or soda. In this manner the valuable nitrogen is again brought back to the soil to enter into the formation of plants.

EFFECTS OF FIRES ON HUMUS.

Fires interrupt this process, which nature has perfected for conserving its energy. They consume all that part of the humus which lies upon the surface of the earth. But they produce changes of more consequence than mere interruption. By the burning the humus is reduced to ash, while the volatile parts and the gases of the combustion pass off as smoke. There are included in these gases of the smoke the carbon or charcoal part of the humus, and other elements, which would have yielded, in connection with this, organic or earthy acids, had decay continued under normal conditions. The most valuable compound, which is in this way destroyed, is the ammonia, or nitrogenous part of the humus. The heat volatilizes this, or sets it free from the complex bodies of which it formed a part, and it is dissipated in the air.

All that the fire leaves is a small quantity of lose ashes. The first wind that comes can blow it entirely away, so far as that locality is concerned. Although they look alike the ash from the pines and from the broad-leaved or deciduous-leaved trees differ much in the proportion of the different alkalies and other compounds of which they are formed. That from the pine contains more of the alkaline ingredients that are soluble in water, almost one-half of their weight being thus soluble, while only about one-sixth of the weight of ordinary hardwood ash can be dissolved in water. If the country is at all hilly the water of the first heavy rain that falls, since it now has no humus to hold it in check, will bear off all the soluble portion of the ash.

A great part of the insoluble ash will in like manner be washed away — if the country is steep and the rain heavy, practically all of it, since, being light and loose, it does not offer the resistance that compact earth does. The soluble parts are decidedly the most valuable. In both pine and hardwood ashes they consist chiefly of sulphuric acid, potash and soda, while the insoluble parts are largely lime, iron and sand in pine ash, and lime with small quantities of magnesia and phosphoric acid in those from hardwoods.

This burning thus robs the burnt lands of their available mineral salts, on which fertility is so largely dependent. If persisted in it eventually greatly lessens the productive power of even the best of soils, especially in hilly or mountainous districts.

It is fortunate that the pine forests of North Carolina, in which so much burning is practiced, grow for the most part in plains and

upon sandy or gravelly soils, which allow ready percolation by
the rain water and afford quick fixation of the alkaline and other
parts of the ash.   By this fixation the soluble part of the ash—and,
as has been said, over 50 per cent. of pine-wood ash is soluble—is
made less soluble, so that drainage water will not bear it off.   This
fixation may or may not affect its availability, so far as the plant
is concerned   About that, as yet, very little is known.

PEATY SOILS.

Moreover, there are soils which are largely made up of humus.
There are peaty or highly organic soils chiefly in the north-eastern
section of the State, around the Dismal swamp, in Hyde and Dare
counties, and areas of smaller extent but similar in character in
Pamlico, Lenoir, Duplin and Bladen counties.   These soils are
either those which from intermixture of sand and their capability of being drained are suitable for tillage or those which
"sour" and are usually too low for complete drainage.   The first
kind frequently rests on a bed of green sand or marl, and before
cleared—for such land is now largely cleared—was covered with
cypress, gums and tulip poplars; much of it, too, was once lakes
that have filled in with *detritus*; while the second kind, which is
"sour," bears a growth of white cedar (juniper) or white bay and
has usually passed through the successive stages of a sphagnum
bog before reaching its present condition.

The sour or peaty lands can be put to no more profitable use than
growing white cedar: keep the fires out, protect the young growth
of white cedar and keep it as nearly pure as possible.   It is
decidedly the most valuable tree of eastern North Carolina;
besides being an abundant seeder and of rapid growth it is exceedingly hardy.   It is a yearly event, however, to hear of a fire, often
originating from carelessness, passing through one of these swamps
and consuming not only the trees both great and small, but frequently burning out the soil until its level is so lowered that white
cedar will no longer grow upon it.   There are thousands of acres
of timbered swamp lands which have been ruined in this way in
the Dismal swamp and in such counties as Dare and Hyde ; and it

frequently happens that peaty lands, even while under cultivation, are burned out and greatly injured.

The State Board of Education controls for the educational fund of North Carolina a large amount of swamp lands in eastern North Carolina. Although some of it is heavily timbered much of it shows to the fullest extent the ravages of fire. Some of this swamp land which remains unsold has naturally a poor, compact soil, and is incapable of sustaining any but an open and scrubby forest growth. Some of it, however, which is now largely covered with cane-brakes, shows conclusively that it was at one time heavily timbered, probably with gums, poplar, oak, and cypress, and that innumerable fires have kept the trees killed down while the cane has spread until now the timber on the lands is perfectly worthless. Considerable areas of these State swamp lands, being regarded as belonging to no one in particular but to the public in general, have been robbed of their valuable timber, and are now annually fired for the purpose of improving the grazing advantages they may offer for the neighborhood herds of cattle. In this way not only has the land been robbed of its supply of mature timber, but the young forest growth also has been destroyed, and thus the future timber supply cut off in large measure.

AREA OF WASTE LANDS IN NORTH CAROLINA.

The census of 1880 gave some interesting and significant figures upon the proportion of the area of this State in woodland, in cultivation, and "other lands," which are idle or unproductive. The amount of this idle or waste land given at that date for North Carolina was 2,000,000 acres, or one-tenth of the entire area of the State was lying out as old fields or other waste lands. An investigation of the forest lands of the eastern counties, undertaken during 1893-1894 by the Survey, showed that there were in the long-leaf pine counties alone, those counties lying south of the Neuse river and east of Montgomery county, over 400,000 acres of sandy forest lands which were producing no valuable forest growth, and under the present negligent management are not likely in the future to do so. This is partly land which, after being lumbered, has been burnt over, or the tree growth, by the many repeated burnings, has gradually been reduced from long-leaf pine forests to scrub-oak thickets, with still a few scattered pines upon it.

The area of the waste lands in other sections of the State has certainly increased considerably in the past fifteen years, so that it can be no exaggeration — in fact, it is a decidedly small estimate — to say that there are now in North Carolina over 3,000,000 acres of waste or unproductive land, including old fields, bramble lands, and forest lands with only a growth of scrub oaks. This makes about one-eighth of the entire area of the State which is at the present time unproductive, either of farm crops or forest products.

AREA OF WASTE LANDS IN EASTERN NORTH CAROLINA.

In the eastern counties, with sandy or sandy-loam soils, and lying within the coastal plain region, the area of the waste land in forty-two counties amounts to over one million acres, or to one-tenth of the total area of those counties. Except the pine barren lands which have already been discussed, any of these lands would be quickly and thickly seeded by the valuable loblolly (short-leaf) pine, and would soon be covered with a heavy young growth of these pines if the fires were kept off. Where the loblolly pine has been lumbered small post and red oaks usually come up. This should be discouraged, as the pine growth is the more valuable of the two. The cause of the oaks appearing is that the young oaks, being able to bear a light shade, are usually found growing under the pines, and if, as usually happens, a leaf-fire passes through the woods after the merchantable pine has been cut, the young pines which come up with the oaks are killed, so that the oaks alone remain. In this section considerably over half of the area is yet woodland, though a great part of it has been lumbered.

AREA OF WASTE LANDS IN THE MIDDLE AND WESTERN COUNTIES.

In the midland counties, which extend westward as far as the foot of the Blue Ridge and eastward to the sandy soils of the low district, the area of the waste lands amounted, according to the

United States Census of 1880, to over 1,500,000 acres, or between one-fifth and one-sixth of the total area. Only a little over one-half of the area of this district is now in woodland. The waste land consists very largely of old fields, many of which have upon them so heavy a sod of broom-grass that it is impossible for a young seedling to gain a foot-hold. Some of these grass lands in the early spring, when the young grass is coming up, after the old has been burnt off, give a scanty pasturage for a few cattle. A large part of the area is typical red clay old fields, washed and gullied, with a poor sod or none, and brambles, persimmon trees and sassafras scattered along the fence rows.

In the mountain counties, except in a few localities, around some of the larger towns and in long settled communities there is very little waste land. In some places, however, there are steep hill-sides which it has been found unprofitable to cultivate, and on which no turf has formed. These consequently have badly washed and gullied. Turf forms on the clayey soils from the slaty rocks less readily than on soils, in similar situations, from the gneissic rocks and schists, and there is, therefore, more washing and gullying of the land on the first mentioned soils. There was in 1880 only one-twentieth of the land in these mountain counties lying out as waste land, neither in fields, pasture nor woodland.

YOUNG GROWTH ON WASTE LANDS.

With the farms scattered as they are through the forests, and surrounded on all sides by numerous kinds of trees it would be almost impossible for a year to pass without the seed from some forest tree being scattered on the numerous old fields and other tracks of outlying, waste or unproductive lands.

Young trees in fact do appear, pines chiefly in the lower division of the State, oaks mixed with pines in the midland division; pines, oaks, tulip-poplar, birches and maples in the mountains, according to the situation. What alone prevents the young trees from growing on the lands are the recurring fires. It makes little difference where situated the waste lands which are connected with farms are almost certain to be broom-grass fields. The burning of this grass, thick and three to four feet high, as it grows, will kill outright the most vigorous young trees four or five feet high or higher. After the grass has been in possession a good many years it forms a sod or turf, if such it can be called, which effectually chokes out the young seedling of all except the hardiest species.

CHANGES IN THE FOREST BODY.

While the changes which our forests are undergoing are gradual they are certainly effective; and the vicissitudes which we know have been occasioned elsewhere by such degenerative changes in the composition of the forest body are none the less sure because slowly produced. These particular changes which are now taking place must be passed through, at least to a certain extent, that we may realize them and take due precaution against their continuance or repetition.

Happily for us the waste of our forests and their alterations have been produced by ourselves in the attempt to satisfy our own needs. Our timber production up to the past decade has been almost entirely for domestic consumption. Primarily a clearing was made for agricultural needs, then the timber was used for building our own houses, fencing our own lands and furnishing our growing demands as fuel. While it was used in such a way, and only the largest specimens culled for special purposes here and there, it was felt that the forests were practically inexhaustible.

INCREASE IN THE LUMBER INDUSTRY.

The past fifteen years, however, have marked grave results. Since 1880 the output of lumber in North Carolina has more than trebled; and the value of the capital engaged in manufacturing has increased from $1,740,000 as reported by the census of 1880 to over $6,000,000 at the present time. It can be said to mark the results of a decade, for far the greater part of the increase has taken place during the past five years. It began along and near the coast where abundant water-ways afforded cheapest transportation facilities for both logs to the mills and the finished material to the consumer; and the timber most in demand has been the hitherto neglected sap loblolly pine. Already this new industry has made deep inroads into the forests of that section. Not content with the so long despised "old-field" pine the swamps have been canalled to reach the cypress and white cedar (juniper). This increased production here is to compensate for the continually increasing shortage in the yield of white pine in the north-eastern States. In Michigan alone this shortage last year was more than the entire annual production of pine lumber in North Carolina amounts to at the present time.

Last year (1894) less than one-half of the seven or eight hundred million feet of lumber manufactured in this State was used here; and each year, as the demand for Southern lumber becomes stronger with the increased consumption and the exhaustion of the forests of the Northwest, a larger proportion of our manufactured product will be for shipment.

The utilization, on a commercial scale, of the hardwoods of this State has scarcely yet begun. Of the 800,000,000 feet of lumber manufactured in this State during 1894 not over 50,000,000 feet were of hardwoods, oaks, ash, walnut and cherry. In spite of the general existing industrial depression the year 1894 chronicled the investment of over $300,000 in the hardwood forests of western North Carolina by northern lumbermen, while there had been previously invested several times that amount. Cypress and yellow pine lands alone, to the extent of 111,000 acres and $346,000 valuation, were owned by Michigan millmen in 1890.

The next decade or two will witness the development of the hardwood industry until it reaches the present proportion of the pine industry; and far-sighted millmen, seeing what an important part the hardwoods of western North Carolina are to play in the lumber market, are attempting to secure them before their value becomes generally known and they increase in price.

Without going further into the future development of the lumber industry in North Carolina, and the markets to be supplied by it, the general facts relative to the extent of our own demands and uses, the yearly rate of accretion of our forests, and their capability of yielding the products, must not be lost sight of.

PRODUCTIVE CAPACITY OF NORTH CAROLINA FORESTS.

There are in North Carolina at the present time about 12,000,000 acres of land in productive forests. The average yield of this land

per annum is only about fifty to fifty-five cubic feet of wood, or
about three-fourths of a cord to the acre. In the mountains, in
places, the yield is twice this amount, and on some of the more fertile sandy loams of the north-eastern counties it is eighty to ninety
cubic feet per annum: but the smaller annual increment from the
greater part of the woodland largely reduces the average.

This small yield is due to the woods being kept open, or non-
productive, by the frequent burnings, and, in the hardwood forests,
by the continual presence of cattle, which keep certain kinds of
young growth eaten down: and of hogs, which never give the seed
of other kinds even an opportunity of germinating. A great part of
the land in the eastern section of the State is kept open by these
agencies, so as to produce only scattering and short-bodied trees.

Some of the second growth woodland in the middle district suf-
fers severely from the same causes, but not so much as in the eastern section, and in many parts of the mountains, especially those
parts where cattle and sheep-raising are leading occupations, the
woods are severely injured by both stock and fires. The harm that
is being done in the mountains is as yet scarcely noticeable, as in
most sections the mature forest still stands unbroken by the axe.
But wherever there is a particular kind of tender growth it is
noticeable that there is, even in the finest forests, no young growth
ready to take the place of the mature trees on their removal. Everywhere destruction is pressing, and nowhere do the natural forces of
growth prevail unchecked.

The average yield per year in this State should be not less than
sixty-five to seventy cubic feet of timber to the acre, and the best
quality of timber at that. With climate as equable as this, and a
rain-fall abundant and evenly distributed throughout the year,
there are encountered none of the natural drawbacks to forest
growth which sylviculturists must so often contend with. The
soils, too, are good, and well adapted to tree-growing; there is but
little land rocky, or excessively dry, most of it, in fact, being deep
and well drained.

If the three million and more acres of idle and waste lands be
considered in this connection as forest lands, and much of it really
is such, the average annual increment is reduced much lower than

is given above. Taking the annual increment on the basis of fifty to sixty cubic feet to the acre, the total yearly yield of our forests will only amount to about 792,000,000 cubic feet, or 9,000,000 cords of wood, while there was taken from the forests in various ways during the year 1894 not less than 975,000,000 cubic feet, or 10,600,000 cords of wood.

There is, then, a great deal more wood being yearly taken from the forests in North Carolina than the yearly growth replaces.

CONSUMPTION OF WOOD IN NORTH CAROLINA.

By far the larger part of the wood used in this State is for domestic fuel. In the consumption of wood for fuel North Carolina ranked, according to the census of 1880, as the fifth State; 7,500,000 cords being used for that purpose alone in 1879 by the people of this State. This is equal to five and one-half cords per head for every person in the State. The value of the wood used for fuel that year was $9,500,000. The consumption has probably greatly increased since then, as the population has increased over 300,000 ; it certainly has not decreased, and is not likely to do so, as wood is not only the fuel for the entire rural population, but is the chief fuel used in all the towns, and in manufactures of all kinds and by many of the railroads now in operation. The general quality of the wood thus utilized is on the whole low, but in the southeastern counties some of the finest pine is being used in this way and the same may be said for young hickory in the midland counties.

Material for construction makes the next largest inroads into the forest. There were at least 200,000,000 cubic feet of wood thus used last year, including lumber, shingles, hewn timber, railroad ties, telegraph poles, etc., and taking into consideration the tops that are left unused in the woods; while there was half as much more used for rail-fencing, fuel for manufacturing and destroyed by fire. Even this does not represent the entire amount of wood that this State furnished, for many million feet of round timber, mostly tulip-poplar, white pine, ash and hemlock-spruce, were sent to mills in east Tennessee by way of the rivers flowing westward out of this State; while 250,000,000 to 300,000,000 feet, board measure, of pine and white cedar logs were exported from the north-eastern counties

to mills in the adjacent parts of Virginia and were there manufactured. Moreover, there were exported from this State over
400,000,000 feet, board measure, of manufactured lumber, besides
square timber, railroad ties, posts, etc., in smaller quantities.

VALUE OF THE ANNUAL WOOD PRODUCTION IN NORTH CAROLINA.

While the single item of fuel is decidedly the most important of
any forest product, both in value and quantity, it represents less than
one-half of the entire aggregate, including both crude timber and
the manufactured product in North Carolina. The value of the
crude timber produced in the State during the past year was about
as follows:

Value of fuel, domestic and for manufacturing...........$10,000,000
Value of saw-logs at mills............................. 3,500,000
Value of round timber, exported........... 1,000,000
Value of railroad ties and hewn timber of all kinds,   500,000
Value of all split fencing, posts, etc................. 500,000

      Total value of all crude products...............$15,500,000

If the value of the manufactured products, including by-products like naval stores, tan barks, various oils and extractives,
be included, the total value of the forest products of this State will
certainly not be less than $22,000,000 and probably will reach as
high as $25,000,000.

The largest items under the head of manufactured forest products are:

Manufactured lumber of all kinds......................$7,300,000
Special industries, veneers and woodenware.   300,000
Cooperage. 125,000
Paper mill products (from pulp), estimated 100,000
Resinous products (naval stores, etc.)................ 1,750,000
Tan barks and extracts................................ 45,000
Wagon, buggy and car factories........................ 600,000
Furniture and repair shops............................ 200,000
Oil of wintergreen.................................... 30,000
Packing boxes, undertaking caskets and agricultural
    implement manufactories........................... 85,000

      Total $10,535,000

If the items which are counted twice be taken from this amount and the value of the crude timber which is not manufactured be added, it will give nearly $23,000,000 as the value of the forest products of North Carolina, and this does not include the great item of building and construction, and several minor items.

This makes the forest crop decidedly one of the most valuable in the State. It is nearly equal in value to the corn crop, and exceeds the combined value of all other grains. It greatly exceeds in value the united products of all the floricultural and horticultural interests.

THE PERIOD REQUIRED FOR FOREST ROTATION.

At first glance it seems paradoxical to state that the yearly production or accretion of our forests is less than 9,000,000 cords, while these same forests are now furnishing over 11,000,000 cords of wood per year. As a matter of fact, the growth which is now being cut up in a decade or two represents the accumulation by the united energies of nature of several hundred years. The only way to ascertain the length of time it requires a forest to reproduce itself is to determine the period required for the individual trees which compose it to reach their maximum age and size.

The loblolly pine (short-leaf or old-field pine of the eastern counties) is under favorable conditions one of the most rapid growing trees. Even it requires fifty to sixty years for the old-field growth, where the soil has been cultivated and enriched, to become large enough for saw-logs with a diameter of about two feet; while the ordinary swamp or rosemary pine growth, which furnishes so much of the timber, represents a growth of over a century. The cypress, which are exceedingly slow growing trees, required two or three times as long to reach their present proportions. Long-leaf pines sixteen and twenty inches in diameter are between one hundred and twenty-five, and two hundred and fifty years old. The mixed pine and hardwood forests of the midland counties represent in their virgin condition from one hundred to one hundred and fifty years' growth; while it will require from two hundred to three hundred years to replace the forests of the mountains when they are destroyed.

3

Taking the figures of the census* of 1880 as a basis, and these figures are doubtless sufficiently correct, about that year the relation between our wooded area, the rate of accretion, and the rate of consumption from, or rather the rate of depletion of our forests, was such that consumption just about equalled the total accretion of that year. Each year after the occurrence of this state of equilibrium there has been a more decided change in the relation of total accretion to consumption; for not only has the forest area been constantly lessening by the encroachment of arable lands, which decreased the productive possibilities, but there has been a constantly accelerating increase in demand as population increased and wants multiplied. We have now reached a situation where depletion largely exceeds accretion.

It is evident that it is from no lack of woodland, for over one-half of the total area of the State is still in trees. Some change, therefore, must be made if the forests are to continue to supply our at present extravagant needs.

FOREST AREA OF NORTH CAROLINA AND ELSEWHERE.

While it has very little bearing on the present subject it may not be inappropriate to give the proportion of the area under forest in some of the principal countries of the world. European Russia has 46 per cent. of its area in woodland, which is about the same proportion which is in this State. Austria has about 32 per cent. of its area in forests. There is 26 per cent. of the German Empire in forests, and these are so well managed that they supply the needs of nearly the entire German people. Systematic forest management and protection of woodland is of long standing in Germany, and the knowledge and experience thus gained has been disseminated throughout all Europe, and is now being utilized in conserving and increasing the efficiency of the forests, not only of the most advanced European countries, but also of India, South Africa and Australia.

In France the system of Germany has been adopted, but with such alterations as best suited local conditions; and by its application

---

*These figures will do for comparison, but for actual calculation the rate of accretion should be considerably smaller, and at the same time the rate of depletion larger, which would make this period of neutrality antedate the year given at least a decade, or about 1870.

the people are largely enabled to supply their own wood with only
16 per cent. of the total area in forest. In spite of the natural economy of the people of France, and the fact that other materials than
wood are extensively used as fuel, considerable amounts of building timber are yearly imported. The forest area, however, is now
being extended by the planting of new forests. The amount to
which coal is used for fuel, and brick or stone for building materials, affects more than any other factors the amount of wood required *per capita*. Omitting the area west of the Blue Ridge, which
is at present heavily timbered, but forms only a small part—less
than one-seventh—of the aggregate area of the State, the wooded
portion of the remaining part is less than one-half of the total area.
Practically all the fuel used in North Carolina is wood, while in
the entire eastern section of the State wood is the only building
material available.

PERMANENCY OF FORESTS.

That one cannot hope to see a young growth reach maturity and
become merchantable sized saw-logs in a few months or a few
years should not deter him from properly protecting and en-
couraging woodland growth. In forests which have been partly
culled, as most of those of this State have been, and where
there are many trees mature or nearing maturity, a thicket of
young growth beneath the higher trees serves an important
function in shading the ground, aiding in the retention of moisture and thus stimulating the older trees into renewed activity,
and especially is this true of oak woods. That tree-growth is most
rapid when the woods are kept open is a saying seemingly based
on fact, but erroneous in itself. The chief requisite in timber is
length of bole and freedom from knots. This is secured only by
means of a thick growth during what is called the height-growth
period of the tree's life. It is under these conditions that the
greatest yield to the acre takes place. Diameter-growth, which is
frequently thought to represent total growth, is attained by full
light conditions; but this thickening of the stem is easily secured
after height is gotten, especially in the hardwood forests where cull-
ing is practiced, by the removal of either the highest trees or those

slightly overtopped, as the individuality of the species may necessitate or economy demand.

Forest growth, both old and young, then, should be regarded as an integral part of the productive wealth of the country. It, however, can only yield returns commensurate with its value when the young growth is carefully protected from fire and the destructive attacks of cattle.

A forest is not a mine of wealth, nor does it represent a mine in any sense of the word. The wealth of a mine is incapable of increase or replenishment when once exhausted. More fittingly, a forest is considered a soil crop, as liable to external injury as any other soil crop, the merchantable trees representing the mature harvest, which should be removed in such a way as to not seriously interfere with the younger growth, which in time will also reach maturity if properly protected.

It must be considered as a permanent factor in wealth production and care taken to husband its resources and to favor in every practical way the growth of those kinds of trees which are now fully recognized as commercially important. Although the forests of other parts of this State have not heretofore been of industrial importance, the forests of the eastern section have during the past thirty years been playing such an important part in its commercial and industrial relations that this fact need not be here dwelt upon. It is sufficient to say that the value of the forest products manufactured in that section last year (1894), from the pines alone, amounted to over $7,000,000, and the very basis of the prosperity of most of the larger towns in that region is the trade in or manufactures of forest products.

The hardwood forests of the middle and western districts are only now beginning to have a general commercial value for their products. This once assured, and the forests then will be commercially as important as in the coast region, and their depletion will be brought about as speedily.

LOSSES FROM FOREST FIRES DURING 1894.

With a view of determining approximately the losses caused by forest fires in the State during 1894 the Geological Survey recently sent a number of circular-letters to persons living in the several counties, and the statements published below have been abstracted from the replies received. Many of the replies were so general or vague in their statements that they furnished no accurate *data*, even for local districts, merely stating that the "damage was large" or "forest fires common." In the middle district, where herds of improved cattle are coming to be kept and are more properly cared for, there are in many of the counties stock or no-fence laws, which restrain cattle from running at large through the woodland. This has caused the practice of burning the woods to be largely discontinued, and very few fires were reported at all from the midland counties. The most numerous and severe fires were in the southeastern counties, where the chief cause for burning was to secure pasturage for a few cattle. Other fires in this section were thought to have originated in efforts to protect turpentine orchards and to protect homes or crops from destruction by accidental fires. Only a few fires were reported as having been caused by the carelessness of railroad hands, or by sparks from locomotives.

SUMMARY OF REPORTS FROM COUNTIES.

Abstracts have been made of the reports received from each county, preserving in each case, where it was possible to do so, the words of the correspondents. The reports are based on answers received from about 250 letters and circulars sent to persons in eighty of the ninety-six counties. The sixteen omitted counties lay in the middle section of the State.

ALEXANDER COUNTY.—Only a few forest fires, and these of the kind that may be called leaf-fires, were reported from this county as having occurred during 1894.

ANSON COUNTY was almost exempt from forest fires during 1894. Besides numerous leaf-fires, a severe fire burnt or seriously damaged the timber on probably 200 acres of long-leaf pine land in the southern part of the county, occasioning only a nominal damage. The destruction of undergrowth among pines by fire is not considered a loss by the people, and in some places the burning of woodland is regarded by them as being rather beneficial than otherwise.

BERTIE COUNTY.—Numerous reports from this county state that while there were many small fires, mostly leaf-fires, the damage done to timber and fencing was not great; 3,000 to 6,000 acres were burned over, with a loss of $1,000 to $2,000. The origin of many of the fires was attributed to railroads. A great deal of young timber was destroyed.

BLADEN COUNTY suffered from a number of large fires during the spring months, burning over hundreds of acres with an almost total loss of timber. One report places the number of acres burned at 5,000, with a loss of $2,000; the largest fire being at the head of Colly creek, and burning 2,000 acres, with a loss of about $600. Firing the woods was intentional, and was done to get tender grass and to protect against unexpected fires in dry seasons.

BRUNSWICK COUNTY.—The fires in this county were for the most part grass-fires, which were set in the spring for pasturage A large portion of the county was burned over; some turpentine timber was destroyed, though its loss did not amount to much. Firing the grass is commonly practiced. It has kept most of the pines killed down and already killed out a large part of the young crop of long-leaf pines, which resulted from the mast of 1892.

BUNCOMBE COUNTY.— Burning the woods has nearly ceased in this county, but is still to some extent practiced in the mountain districts, where cattle are grazed in the woods.

BURKE COUNTY.—There were 7,000 to 8,000 acres of timbered lands reported as burnt over in the South mountains alone, killing a large amount of pine timber and burning much fencing. Fires also occurred along the Blue Ridge and along the line of the Western North Carolina railroad, but did no great damage. All forest fires after the first of March kill much of the young forest growth. Fires were said to have originated from burning brush, o'possum hunters, and more frequently from incendiarism; those along the railroad from locomotives, and some were set by chestnut gatherers. One correspondent thinks that burning the dead herbage and undergrowth does good by killing insects; but sometimes it also kills yellow pines and growing timber. The benefit that may be done in the way of killing insects is doubtless insignificant as compared with the damage resulting from these forest fires in the way of killing out the young tree growth.

CALDWELL COUNTY.—One correspondent states that he knew of nine large fires; one in March in Lenoir township, one in March in Globe township, one in March in Patterson township, one in April in Yadkin Valley township, one in November in Patterson township, and another there in December; one in December in Yadkin Valley township, one in Lower Creek township and one in Kings Creek township. There were from 40,000 to 15,000 acres burned over, with a loss of about $5,000.

Another correspondent mentions seven fires in his section of Caldwell county during the year 1894. One of these was in June in Globe township, and in August and again in November in the same township. In November, in the central part of the county, near Lenoir, two fires occurred, one in Yadkin Valley township and another in Yadkin Valley and Patterson townships in December. There were about 40,000 acres burned over, with a damage to timber of about $6,000. Young poplars and chestnuts suffer most from these fires, but white pines are very much injured, oaks are scarred and often injured for lumber, as is also the case with chestnuts and hickories. These fires, this correspondent thinks, will cause the gradual disappearance of the chestnut. The trees are scorched by the fires and decay sets in on the burnt side. Fires are set in the woods to make the grass grow for cattle, and to burn the leaves so hogs can get mast. Wherever the stock-law has been introduced the number of fires has been much lessened.

CATAWBA COUNTY.—The damage to timber from forest fires in Catawba county, as in many of the other counties in the middle section of the State, was slight so far as could be learned. The timber in the county is largely of hardwoods, usually with hardwood undergrowth in the forests.

The damage to timber in MECKLENBURG, DAVIDSON, RANDOLPH, GUILFORD and PERSON counties was caused by occasional leaf-fires, and was on the whole not great.

CAMDEN COUNTY.—A forest fire on the "lake" side burned through to the swamp, near the Currituck county line. Over 2,000 acres were burned over, with a damage of $2,000 to timber and considerable damage to other property. One report states that a large fire usually occurs in this locality every year. There were several

brush and leaf-fires in other parts of the county, which occasioned
no more serious loss than the destruction of the young growth.

COLUMBUS COUNTY.—Correspondents stated that burnings had
occurred in numerous localities in this county, but they were unable to give the extent of the losses. One fire, burning over about
500 acres, occurred near Fair Bluff, and occasioned a loss of $125.
It is said to have been due to the carelessness of railroad employees.
Other and larger fires occurred in the north-western and western
parts of the county, near Hub and Lennon's cross roads. The large
turpentine forests of this county have been gradually destroyed by
fires, which have also burnt the young growth down as fast as it has
appeared.

CUMBERLAND COUNTY.—A correspondent living in the eastern
section of the county writes: "It is impossible even to approximate
the number of forest fires occurring in this county. They take
place mostly in the late fall, when their illumination is almost
nightly seen and frequently at every point of the compass at the
same time. The damage caused by them during the past year was
very great, especially to old pines, and the area burned over large.
The greater part of the young growth in the forest has been destroyed
except the black-jacks, which in their green state are nearly fire-
proof, and it is fortunate for the people of the sand-hills that it is
so. Burning the woods is far too common, but is clandestinely
done, as public sentiment decries it. It is done to stimulate early
growth for worthless cattle."

EDGECOMBE COUNTY.—A correspondent writes that so far as he
was able to judge, after consultation with others, more than two-
thirds of the entire forest area of this county was burned over in
February, March and April. Fires in the woods at other seasons
occasion but little damage. The damage to the standing timber
each year is from 5 to 10 per cent. of its value on the area burned
over. When a forest is not burnt over in a number of years the
damage then from fire is much greater. Losses from burning of
fences amounted in 1894 to about $500. Fires are usually acci-
dental from burning old fields and brush and hunting with torch.
They have burned down most of the boxed pines and destroyed the
young growth of timber.

On the basis above given 60,000 to 80,000 acres must have been burnt over, with a loss of $10,000 to $20,000.

GRAHAM COUNTY.—Burning the woods has been practiced in this and in Cherokee county ever since they were settled, and before that time the Indians practiced it. The trees in many places, especially the chestnuts, have been scorched on one side and then hollowed out from the effects of the fires. Much other timber and young growth is injured. Many of the mountains in Graham and Swain counties were burned over by the Indians during the past year. It is safe to say that one-fourth of the mountain lands of these three counties, Graham, Swain and Cherokee, was burnt over during the past year.

HALIFAX COUNTY.—A destructive fire occurred in March just west of Enfield, which burned over some 4,000 or 5,000 acres of timbered land. The damage, the correspondent thinks, would be at least $1.50 per acre, all undergrowth less than three inches in diameter being killed. Fires in the woods in this county are usually traceable to camp-fires, hunters and tramps. No reports were received from other sections of this county.

HARNETT COUNTY.—Leaf fires, or grass fires, set to better the pasturage, burned over a large part of the wooded area of the county lying south of Upper Little river. This part of the county has been burned over so many previous times that nearly all the timber on it has been destroyed. There were a few unimportant fires in the northern part of the county.

HENDERSON COUNTY.—One report states that a large part of the forest lands, at least one-third, was burned over during the winter of 1893–'94, between November and May, with a heavy loss of timber. The same report states that at least two-thirds of the standing timber has been damaged by fires which occur regularly each season, and which are purposely started to better the pasturage. Some fires, however, are accidental.

JACKSON COUNTY.—The outside mountain lands, or wild lands, are yearly burned over to supply grazing. At least a third of the area of these lands was estimated to have been burned during the past year. Great damage is done to the poplar and chestnut timber; indeed it is difficult to find in these wild lands a tree of these kinds that is not defective at the base from this cause.

JOHNSTON COUNTY, like the other counties in which the long-leaf pine predominates, yearly has a large part of its area burned over for the pasturage. Besides all the young pines being destroyed the damage to timber last year was estimated to amount to five per cent. of the value of the timber on the area burned over, and this was very little in excess of the usual annual loss.

JONES COUNTY.—Only a few local fires were reported from this county.

MACON COUNTY, like so many of the other mountain counties, yearly has a large part of the "wild lands" burned over. And although the fires are chiefly leaf-fires they have caused great damage to the timber. Between 10,000 and 20,000 acres were estimated to have been burned over during the past year. The loss from the fencing destroyed was placed at more than $2,000.

MADISON COUNTY.—Although there were several fires at various places in the county there was only a single destructive one reported, which burned over about fifty acres. Burning the woods is practiced in many sections of the county to keep the woods open and better the grazing.

MITCHELL COUNTY.—Thousands of acres, mostly on southern slopes, were reported as burned over during the past year in this county. One correspondent says that although the damage to standing timber from a single fire appears to be small the continual burning, year after year, results in serious damage, killing much of the timber and seriously injuring the rest, so that its value has been lessened one-half by the mere repetition of the leaf-fires. On many south mountain slopes many of the larger trees have been destroyed and only a brushy growth occupies their place. The practice of burning the woods for improving pasturage is a common one in parts of this county.

Many of the statements made about the practice of firing and the resultant damage to the woodland of Mitchell county will apply as well to parts of the adjoining counties of Yancey and Watauga.

MONTGOMERY COUNTY.—Besides the usual spring fires set for bettering pasturage a few destructive fires were reported to have occurred in the wake of the lumbermen in the eastern and southern parts of the county.

MOORE COUNTY.—The forests of this county have been devastated in the past few years by several large and destructive fires. One in 1892 consumed a large amount of timber in the southern part of the county and wiped the village of West End out of existence. Another in 1893 destroyed long-leaf pine timber to the value of $50,000. Fortunately, however, there were no destructive fires during the past year, although a large portion of the county was burned over for pasturage, or to protect property against an unexpected fire in a dry season.

NEW HANOVER COUNTY.—There were between 3,000 and 4,000 acres in this county, covered with young pine trees, burned over. Many of the young trees were killed; otherwise the damage was not great. Fires have greatly decreased in this county since the adoption of the stock law.

NORTHAMPTON COUNTY.—Only a few local fires were reported from this county.

ONSLOW COUNTY was visited by several destructive forest fires. One correspondent mentions five large fires which occurred on the west side of New river in the turpentine woods. These burned over 1,000 acres, with a loss of over $6,000, mostly of pine timber and young growth. Another person writing from the southern part of the county says that one fire in the neighborhood of Brown sound burned over about 8,000 acres, with a loss of $5,000. This fire originated from lightning. Another fire was mentioned near the Duplin county line which burned 300 acres. Burning is commonly practiced in this county for improving pasturage.

PERQUIMANS COUNTY.—Six fires occurred in April and May in Parkville and New Hope townships, burning over a large area. While the damage to standing timber was not large a great injury was inflicted on the underwood, which was all killed. These fires were purposely started, and are thought to improve the forests and drive out foxes and other wild animals. The practice of burning is here frequent.

RICHMOND COUNTY.—There were 6,000 to 7,000 acres of timbered land burned over in this county, the burned areas lying for the most part in the southern and eastern portions of the county. The most extensive one was in Marks Creek township. This burned over 1,000 acres of recently lumbered pine lands, killing all the young pines and a great deal of scrub oak timber. The wire-grass in the pine-barrens is fired every spring to better the grazing and to kill out the young growth.

ROBESON COUNTY seems to have suffered less than usual from fires during 1894. There were a great many small grass-fires which destroyed the undergrowth. In that portion of the county where the stock law is in operation there were very few fires.

SAMPSON COUNTY.—A correspondent from this county, which lies in the long-leaf pine region, estimated that there were 100,000 acres of timbered land burned over, with a loss of $50,000. The loss was mostly in young growth, to a less extent in timber. Burning is practiced for pasturage; but some fires are accidental. Most of the young pine growth in the forests, especially in the southern part of the county, is kept killed down. Another correspondent in the south-western part of the county states that there were no serious forest fires in his section during the past year, though there were many smaller ones.

WAKE COUNTY.—There is a considerable part of this county burned over every fall and spring, damaging a great deal of young growth. The same applies also to the adjoining counties of Nash and Chatham. Most of these fires are purposely started or escape from brush-fires where new ground is being cleared.

WAYNE COUNTY.—A large part of the southern section of this county, where it is very sandy and grassy, was burned over in the spring of 1894 for pasturage. Fires have killed all of the young long-leaf pines in this part of the county and only scrub oaks have taken their place.

REMAINING COUNTIES.—No reports, or none that were satisfactory, were obtained from Tyrrell, Washington, Dare, Hyde, Duplin, and Carteret counties, lying in the eastern section of the State. These counties, however, have a large proportion of their areas under swamp.

No reports were received from Wilkes, Stokes, Polk and Rutherford counties; but it is safe to say that a large part of the woodland in them was burned over, as was the case with the other piedmont counties of Burke and Caldwell.

No attempt was made to secure any specific information from the counties in the middle portion of the State, as these have hardwoods for a large part of their timbers, are thickly settled and the woodland is not so much injured by fires as elsewhere in the State, nor is the practice of burning so frequently resorted to in order to stimulate a scanty pasturage. There is still room, however, for a large reform.

AGGREGATE VALUE OF THE PROPERTY DESTROYED BY FOREST FIRES.

The difficulty of getting any reliable figures concerning forest fires is well shown in the incomplete reports from the several counties enumerated above. The nature of the subject explains the difficulty of obtaining full statistics. The large number of the fires—most of them being leaf-fires, which are of so frequent and general occurrence as to fix neither locality nor extent definitely in the memory of even those who saw them—and the inability of the most widely informed persons to estimate and report accurately beyond a local district, render the accumulation of exhaustive evidence upon the extent and destruction of these fires well-nigh impossible.

The damage as stated by the correspondents from the several reporting counties must have aggregated over $400,000, and there must have been between 800,000 and 1,200,000 acres burned over during the year. The damage attributed to Sampson county may be in excess of the actual losses sustained ; but that from the other counties is in all cases probably underestimated, usually a third or a half smaller than in reality. Moreover, it is difficult to fix any standard by which losses can be ascertained ; for only mature trees of certain merchantable species are considered in making the estimates, while the destruction of kinds with no commercial valuation as yet, and young growth, is counted as nothing. The counties reporting, too, embraced only about one-half of the area of the State, and the writer from his own observation of the damage wrought in previous years in these non-reporting counties would estimate the damage of them at over one-half of what it is in the other counties, or over $200,000. The entire loss in 1894 caused by forest fires in the State was certainly not less than $600,000; and from 1,500,000 to 2,000,000 acres of forest and waste lands were burned over. Only on a comparatively small part of this land, however, did the fires amount to more than leaf, or brush, or grass-fires; but, as has been shown already these result ultimately in the total destruction of the forest.

THE RELATIVE PREVALENCE OF FIRES IN DIFFERENT REGIONS.

The most numerous and severe fires in this State occur in the south-eastern counties in what is known as the long-leaf pine belt, or pine-barrens. High and thick grasses cover the ground and when dry in winter and spring form a fuel which carries fire before a wind at an alarmingly rapid rate. New Hanover county and parts of Robeson and Pender, where the stock laws are in force, have comparatively few forest fires. In the north-eastern counties there are neither so many fires nor is the damage resulting so great as it is farther south. Almost one-fourth of the definitely reported fires occurred in the south-eastern counties. So difficult is it to extinguish these wire-grass fires when once well under way that they have been known to burn from Hamlet to Fayetteville, a distance of forty miles.

Next in order, as far as number and extent of area burned, come the south-western mountain counties. The soil here is in many places in a condition in which it holds but little water, so the dry leaves burn well, and, wherever there are any Indians, the woods are regularly burned; but the Indians are by no means the only offenders. The mountainous parts of the piedmont counties suffer a great deal also, the ridges being steep, and much pine mixed with the hardwoods, so that a fire once started in a dry season burns briskly. The woods of the other counties west of the Blue Ridge are frequently burned when the season is dry enough; but the conditions in these are not as conducive to fires as in the south-western counties. The fires in the piedmont and mountain counties are leaf or brush-fires, rarely damaging directly anything except the bases of the trees.

The midland counties enjoy a comparative exemption, no very large fires being reported from this section. The people are beginning to realize their damage and prevent them and make every effort towards extinguishing them when once started. They rarely pass beyond brush or leaf-fires.

The fires in the eastern part of the State not infrequently pass from the grass to the pine trees which have been boxed and either ruin the face of the turpentine box or burn the boxes out so that the first storm will blow the trees down. After the face of a box has been burned only a low grade of rosin can be obtained from it, since the cinder darkens the resin.

THE ORIGINS OF FOREST FIRES.

By far the greater number of the fires, at least two-thirds of them, seem to be of intentional origin. And at least two-thirds of those purposely set are to secure or improve the pasturage. In the eastern part of the State these spring fires burn off the tough and thick old growth of wire-grass and broom-straw and the cattle can very early in the spring for several weeks get a fairly good pasturage. The grass, however, soon becomes too hard to be eaten. In the mountains fires are set to get rid of the leaves, so that the young grass can be easily reached in the spring; to burn off the stiff weeds, etc., and, what is much more important to the grazer, keep the young tree growth killed down. Keeping the young tree growth killed down exercises a twofold influence: it keeps the woods open so that grasses and herbage can grow, for these will not grow where it is too shady; and it keeps all pines and other conifers killed down, as these do not so readily sprout from the stump; while keeping them killed keeps the hardwoods, which cattle eat, of low growth, or always sprouting from the roots, so that they afford young and tender shoots within easy reach of cattle. It must be borne in mind, however, that these fires also destroy much of the grass and other annual and perennial herbs and shrubs, by destroying both the seeds and the plants themselves, in the forests and about the margins; and that in this way in the long run the pasturage in the forests is injured rather than improved by these repeated burnings.

Burning to protect houses, etc., is said to be a frequent cause for firing in the south-eastern counties. These are fires set on still days to get rid of the inflammable material, so that there will be no danger to farms and crops and houses from a chance fire in windy weather. A few fires were reported as set to enable hogs to find mast; and some by chestnut hunters; some from malice. In the turpentine orchards they are intentionally set to keep the growth down and get rid of the inflamable grass before the boxes are cut, or the sap begins to rise in the trees. Some were said to be set to get rid of insects, pine borers, etc., which is certainly using a very dangerous remedy for an insignificant evil.

Still other fires were reported as being started to drive game from cover. Most of the fires in the eastern and many in the western part of the State are started by indigent persons who are amenable to no law, who regard all property as open to destruction and forests as communal property: persons whose parents were hunters and who themselves are scarcely yet seriously affected by the civilization which defines property and allows to the individual its possession. The few fires that were reported as being of accidental origin were from hunters at night, campers, locomotives, lightning, and many from burning brush or logs in clearing land.

THE FOREST FIRE LAWS.

The general law in North Carolina relative to forest fires has remained on the statute books practically unchanged, and largely a dead letter, since it was enacted in 1777: and during this nearly a century and a quarter that has passed since that time, fire has destroyed more timber in the State than the lumberman has cut. The law is as follows:

"No person shall set fire to any woods, except it be his own property; nor in that case, without first giving notice in writing to all persons owning lands adjoining to the woodlands intended to be fired, at least two days before the time of firing such woods, and also taking effectual care to extinguish such fire before it shall reach any vacant or patented lands near to or adjoining the lands so fired."*

"Every person wilfully offending against the preceding section shall, for every such offense, forfeit and pay to any person who will

*Code, 1883, c. 7, s. 52.

sue for the same fifty dollars, and be liable to any one injured in an action, and shall moreover be guilty of a misdemeanor."*

Besides these laws relating to firing the woods there is the following one in regard to wagons and camps:

"If any wagoner or other person camping in the open air shall leave his camp without totally extinguishing his camp-fire he shall be liable to a penalty of ten dollars, to be recovered by any person suing for the same, and shall furthermore be liable for the full amount of damages that any individual may sustain by reason of any fire getting out from said camp, to be recovered by action in the Superior Court for the county in which said camp may be situated, or in which said damage may be done: *Provided*, that this section shall apply only to the counties of Cumberland, Harnett, Bladen, Moore, Hertford and Chowan"†.

The last section was passed in 1864–'65, is in operation only in six counties, and covers merely the case of accidental fires from campers. A statute similar in its provisions but imposing a heavier fine was passed in 1885. Its application extends only to thirteen counties, embracing most of those in the south-eastern part of the State. It is as follows:

"If any wagoner or other person camping in the open air shall leave his camp without totally extinguishing the camp-fire he shall be guilty of a misdemeanor, and on conviction fined not exceeding fifty dollars or imprisoned not exceeding thirty days, at the discretion of court, and also be liable to parties injured: *Provided*, that this statute apply only to the counties of Onslow, Pender, Edgecombe, Robeson, Wayne, Columbus, Cumberland, New Hanover, Bertie, Cabarrus, Harnett, McDowell and Davie."‡

There is besides the above a special law applying only to Pamlico county which was passed in 1889. This forbids any person from firing his woodland or marsh-land between May 1st and December 31st, unless it be separated by a swamp or stream from the lands of other persons.§

*Code, 1883, c. 7, s. 53.
†Code, 1883, c. 7, s. 54.
‡Laws of North Carolina, 1885, c. 226, s. 1.
§Laws of North Carolina, 1889, c. 225, s. 1.

There is no doubt that section 52 of the Code of 1883, in regard to notice of intention to burn being sent to owners of adjoining woodland, is a dead letter. Over two-thirds of the woodland fires in this State are purposely set, and their annual number must reach at least 500 to 700. The writer is unable to learn of a single case in which such notice was given by the parties who started the fires. Although a large number of the fires annually occurring are traced to night hunters and a few to day hunters, there is no statute providing for fires originating either through carelessness or design of persons so engaged in hunting.

VARIOUS VIEWS UPON THE EXISTING FIRE LAWS.

A question was asked many persons concerning the efficiency of the existing forest-fire law; and for suggestions by which its efficiency could be improved or its provisions more effectually enforced. As all shades of belief and criticism found expression some of the more prominent and pertinent answers are given below.

Over one-half of the answers received indicated that their writers were ignorant of the existence of a law concerning forest fires or burning brush, etc. The chief idea expressed was that if a person's timber is seriously damaged by a fire started on his land by another person a suit can be brought for damages against the offender.

Such action is brought against railroad companies when fires are traceable to locomotives, and the assessed damage is usually paid in such cases.

A prominent correspondent in one of the piedmont counties said: "A law to prevent them (fires) would be difficult to enforce from the fact that it would be impossible to find out who does the firing."

One in eastern North Carolina wrote: "The no-fence law would prove the most efficacious in regard to those of intentional origin. Penalties should be enforced on the responsible parties when resulting from carelessness."

A writer in one of the south-western counties said: "I do not think the present law strong enough. Fires can only be prevented by passing strict laws with heavy penalties, and the strict enforcement of the same by the courts, which has not been done heretofore in this part of the State. Most of our people would be glad to see the law strictly enforced and a stop put to forest fires, as nothing is doing this part of the State more harm."

Another states: "We have plenty of law, but it is difficult to enforce it. The people need to be educated on the subject. Wherever the stock law has been introduced it has been found that the number of fires has been much reduced."

The writer of the following, like many others, appears ignorant of the existence of a fire law: "If a fire gets out on another's land not only should the party offending have to pay the damages but be subject to a suitable penalty for misdemeanor."

A person in one of the south-eastern counties whose lands, he says, have been much damaged by fires thinks it would be a good idea, in neighborhoods where fires are common, to have a man to look after them and see that offenders are reported and punished.

INFLUENCE OF LUMBERMEN IN CHECKING FIRES.

The influence that earnest millmen can exert, if so minded, will probably have more effect in their respective localities than the enactment of any law, no matter how severe and exacting may be its provisions. If these millmen, who form in reality one of the most interested classes, will take the matter in hand and make their employés understand that the protection of the young growth from fires is the assurance of labor for them, and it is the only way that the lumber industry can be perpetuated in this State, there will be gained a strong position of vantage. These laborers and the opinions held by them reach a class which writing of no sort can reach or influence; and it is this class which, either ignoring or neglecting to consider both the moral and pecuniary aspects of their acts, is responsible for far the greater number of the forest fires. If once the 15,000 men engaged in handling lumber and timber in this State are made to understand the advantage of protecting young forest growth and preventing fires, both by not setting them and by informing against those who do, public sentiment will come to their support, and we will begin to realize that the forests of the State may have a future as well as a present.

Although it is the lumbermen whom the fires—especially in the eastern part of the State—are most injuring and who would be the most benefited by their suppression, many of them express

absolute indifference to any effort to mitigate the evil or to show
the great loss occasioned by burning. They even appear to oppose
any reform, and regard with decided disfavor any effort towards
securing it. There are many millmen, however, who have aided
in securing valuable *data*, and who express their readiness to
co-operate in any measure which assures any abatement as to the
number and extent of the fires.

RELATION OF THE BURNER TO THE FORESTS.

The turpentine workers also regard it as right to burn the grass
and undergrowth in turpentine orchards, and in spite of the fact
that the fire is liable to extend to the trees continue thus to fire the
grass. It is impossible, however, to secure the boxed trees and the
highly inflammable scrape-covered faces to the boxes as long as
the undergrowth is burned; and it is equally as impossible to
secure a regrowth of long-leaf pine as long as the burning grass
consumes all seedlings and seed.

There are in the south-eastern counties clearly two classes who
are interested in the burning: (1) the timber owner who sees his
woodland yearly deteriorating in value at the hands of another,
and who, it seems, can obtain no redress for his loss—cannot
even secure a suppression of the agency of destruction ; and (2) the
person who for the immediate benefit he fancies is derived from
the act innocently or willfully, directly or through gross neglect,
burns off the lands and destroys the timber or other property
belonging to another citizen.

These two elements seem irreconcilable. As a matter of fact,
however, their aims and dependencies are similar. It is apparent
that the person who does the burning does not realize the relation
between himself and the woodland. For in many cases, and as a
general rule in the south-eastern counties of the State, a great part
of his existence is dependent upon it, and that the more forest
there is the greater will be his benefit, his cow or the ease with
which he can work his rented turpentine boxes notwithstanding.

INFLUENCE OF THE NO-FENCE LAW IN CHECKING FIRES.

In a few places in eastern North Carolina the stock or no-fence
law has been tried for a number of years; and, though the object

in securing it was solely for the improvement of cattle and to lessen the cost of fencing, it has produced, in those localities where it has been tried, a decided change in the appearance of the young growth in the forest, and aided to lessen the number of fires by removing in a large measure the incentive to burning. To show the relative importance of protecting the forests to the pasturage gained by turning stock into the forest and firing the forest on the supposition that this improves the pasturage, it will suffice to state that the total value of all the cows in Moore county would not pay for the timber destroyed in that county alone during the years 1892 and 1893 by two fires. Harnett, Richmond and Bladen counties all show a similar state of affairs.

And not only would the forest fires become less frequent if the stock or no-fence law be generally introduced, but the provisions of the law quoted above in regard to the punishment of persons for firing should be rigidly enforced; and if, as the law now stands, it cannot be made operative it should be judiciously amended and provision made for its proper enforcement.

THE MAINE LAW IN REGARD TO FOREST FIRES.

The State of Maine presents conditions strikingly analogous to those in North Carolina. As is well known it is a great lumbering State. A people almost entirely dependent on agriculture and forest products for their prosperity, and whose manufacturing had never assumed importance, they finally recognized the interdependence of the agricultural and the forestal interests, and to secure the estoppage of fires and the unreasonable waste of timber, which goes on where it is abundant, adopted a series of laws relating to the suppression of fires. Their laws empowered a commissioner with the right to publish rules and to organize a corps of fire wardens and, finally, to bring suit against offenders to recover damages for loss sustained.

This law, being the first of its kind in the United States, is of sufficient importance to be summarized. The following excerpts and condensations are from the law as published in the first report of the Forest Commission of the State of Maine, 1891, which was obtained through the courtesy of Mr. Charles E. Oak, the present land agent of the State. While the provisions of this law are at present inapplicable to the conditions in North Carolina, they will serve to show what stress another State lays upon the strict enforcement of the laws relating to forest fires.

The law provided that the land agent—the State of Maine having large areas of forest lands for sale—should be forest commissioner. It is his duty to collect statistics about the forest resources and products of the State, and to receive the reports of the fire wardens. The selectmen of all towns are fire wardens, and so also are designated other persons living in various parts of the counties where fires occur. These latter are appointed by the county commissioners. It is the duty of the fire wardens to call out citizens in case of a fire to aid them in extinguishing the fire or controlling it; and to have the authority of deputy sheriffs to force persons to help to extinguish a fire. All persons not answering their summons are subject to a fine. The area burned and value of the property destroyed must be reported to the commissioner. It is the duty of municipal officers in towns, and of county commissioners with respect to other localities in their counties, to proceed immediately into a strict inquiry into the cause and origin of fires within woodland, and in all cases where the fires originate from unlawful acts to cause the offender to be prosecuted without delay. There are provisions concerning the clearing of rights of way by railroads, and the use of spark-arresters and other precautions; and provisions about hunters, campers, etc.

In a private letter the present commissioner states that the law is very satisfactory in its workings, and has greatly lessened both the number of fires and the losses caused by them. Its provisions, he says, are carried out and offenders brought to justice. It has, moreover, awakened a new spirit among the people—that of protecting young forest growth instead of destroying it.

LAWS IN OTHER STATES.

The Adirondack lands, belonging to the State of New York, are under a system of laws more strict in detail but similar in outline to those in force in Maine. The aim of the forestry commission legislation in New York is to restrict or supervise cutting of timber on State lands and to prevent fires occurring on the Adirondack mountains. The people of Pennsylvania, as well as those of Vermont, are taking an active interest in the protection of young growth and the prevention of its destruction by fires and cattle. Fires in Pennsylvania annually occasion a great loss of timber in the mountainous parts of that State, and that State now has a forest commission whose duty it is to disseminate information concerning the proper care and protection of forests, and to show the injurious effects of fires in woodland.

INDEX.

ERRATUM.

Page 25, line 11 from top, read *Ailanthus* for *Ailianthus*.

www.ingramcontent.com/pod-product-compliance
Lightning Source LLC
Chambersburg PA
CBHW022006190326
41519CB00010B/1401